May this "Fan North" Bouquet warm your heart with Peace + Joy!

Melvin Stephen Lord
Oct 2005

Dedication

Lovingly dedicated to my family: Jim, my husband of 51 years; Our daughter Wendalyn, her boys Phillip and Jacob; Our son Wayne; Our daughter Joanna, her husband Dean, their children Deanna, Joleena, and David; Our son Jim, his wife Connie and their girls Catherine and Jenna.

And to our great God of love and grace, who brought into being my dear family, wildflowers, and the thoughts for this book.

Acknowledgments

Hugs and sincere thanks to: Gregory Sizemore—my teenage friend who enjoys wildflowers and understands computers. Thanks for arranging the manuscript in order and onto a disk; Wendalyn Tisland and Joanna Baugh—appreciate your taking time from busy schedules to proofread for me; Stacy Urich at the Wynn Nature Center, Homer—for help in identifying so many wildflowers; Lura Nelson—for being my special friend and encourager.

A Bouquet From Alaska
Wildflower Devotions from the Land of the Midnight Sun

Melva Stephens Lard

Copyright 2003 by Melva Stephens Lard
—First Edition—
ISBN 1-59433-002-6
Library of Congress Catalog Card Number: 2003110276

All rights reserved, including the right of reproduction in any form, or by any mechanical or electronic means including photocopying or recording, or by any information storage or retrieval system, in whole or in part in any form, and in any case not without the written permission of the author and publisher.

PO Box 221974 Anchorage, Alaska 99522-1974

Printed in Korea

Contents
An Alaska Bouquet

Introduction			3
Strawberry Spinach	The Bible	Fairbanks	5
Pineapple Weed	Friendship	Fairbanks	7
Chickweed	Healing of Memories	Fairbanks	9
Twin Flower	To be like Jesus	Fox Spring	11
Fireweed	Growth in Hard Times	Fairbanks	13
Dwarf Dogwood	The Cross	Nenana	15
Arctic Lupine	Spiritual Radar	Clear	17
Pasque Flower	Winter of the Soul	Clear	19
Siberian Aster	Straight Under Pressure	Healy	21
Watermelon Berry	Clouds	Denali Park	23
Arctic Daisy	Unconditional Love	Denali Park	25
Butter and Eggs	Breakfast for the Soul	Wasilla	27
Monkshood	Balance	Wasilla	29
Rattlebox	The Tongue	Wasilla	31
Grass of Parnassus	Awe of Creation	Wasilla	33
Ladies Tresses	Believers are Precious	Wasilla	35
Mountain Harebell	Spiritual Intimacy	Hatcher Pass	37
False Hellebore	Fear of Being False	Hatcher Pass	39
Blue Flag	Be Transplantable	Anchorage	41
Alaska Cotton	The Garment of Praise	Anchorage	43
Blue Columbine	Grace	Ninilchik	45
Beach Fleabane	Home in the Heart	Homer	47
Willow Herb	Equality of Gifts	Homer	49
Yellow Monkeyflower	Laughter	Homer	51
Goldenrod	Heaven	Homer	53
Yellow Paintbrush	Love	Homer	55
Pink Pyrola	Dwelling Places	Homer	57
Forget-Me-Not	Grow Up	Homer	59
Nettle	Two Natures	Homer	61
Wild Geranium	The Family	Homer	63
Alaska Map	Wildflower Field Trip Route		64

Introduction

Alaska is a great land of snow covered mountain peaks, glaciers glistening in the sun, vast spaces of tundra and wilderness. Surely there is no better place to hear the music of God's heart than thru His creation. From the pages of the Bible, Psalms and Isaiah speak of mountains that break forth in singing, while forests add their song and trees clap their hands with praise. Listen quietly and your heart will hear it!

It is my joy to share another of Alaska's creative wonders, over-looked by those who desire to see only the gigantic. This book will acquaint you with a tiny part of God's magnificent creation, a part that also declares His mighty work, praises Him, and causes the heart to sing. Wildflowers are that creation, liberally sprinkled over the state. They are adorned in beauty greater than the robes of King Solomon himself.

Our years in Alaska were filled with many responsibilities and little time to pursue my love of flowers, especially those planted by God Himself. However, retirement in rural Alabama gave me the time to become a serious student. Unexpectedly, our gracious God stepped into my studies a few years ago and enable me to see wildflowers with spiritual eyes. He began to reveal, in each flower, lessons about the character of God, the believer's position and walk with Him, and our relationship with other Christians. I started to write devotions on these thoughts, thinking it would be a special legacy to give to my adult children (little did I dream that the Lord wanted a bunch of other folks to read them too!) I felt so unworthy and unqualified for His use, but He just needed a willing heart so I gave Him mine. Then He met me where I was and used me as He found me! What grace, Amazing Grace!

Join me now in an exciting adventure. Twenty-five years and many summers qualify me to be your guide during our unique "ARMCHAIR WILDFLOWER FIELD TRIP." Seat yourself comfortably, perhaps with a cup of coffee or tea. Prepare the vehicle of your mind to carry you from stop to stop. Remember, we are not looking UP for the huge and enormous, rather DOWN, at our feet, for the small and minute.

Follow the map (page 64) as we begin in Fairbanks, go north ten miles to the Fox Spring and then work our way south, past Anchorage, and conclude our trip in Homer.

Blessings on you as you travel.
Ready? Let's Go!

Strawberry Spinach
Chenopodium capitalum

By August, the edges of the lawn and flowerbeds in my daughters yard in Fairbanks usually provide an ample supply of Strawberry Spinach. The flowers of this unique plant are very tiny. I don't recall ever finding Strawberry Spinach in bloom, but it becomes very noticeable as the large reddish fruit develops.

After years of "knowing" that it was edible (my book knowledge), I decided to give it a taste test to determine for myself the credibility of its name. Sure enough, the leaves and stems are very similar in taste to its close relative, spinach. The berries have a slight strawberry taste eaten raw, and make a beautiful, delicious syrup.

Strawberry Spinach reminds me to eat healthy spiritually. Popeye the Sailor knew the health benefits of a can of vitamin-rich spinach. Whenever he needed extra strength, he'd pop open a can and gulp it down. Whenever I need extra spiritual strength, I "pop open" God's Word and nourish my soul. How sweet are its promises, precepts and principles.

As I taste the sweetness of His Word, I join with the psalmist, David, when he said, "Oh, taste and see that the Lord is good; blessed is the man who trusteth in him." (*Psalms 34:8*)

We may hear from others that God's Word will provide strength in time of need, comfort in sorrow, wisdom in uncertainness, joy amid trials, and peace in confusion and hope in desperation. We will only KNOW this to be true as we read God's Word and apply it to our daily lives.

Father God, forgive me for the times I've attempted to experience Your Word vicariously. And Father, a simple taste is no longer enough. I desire to set at Your table and FEAST!

That you may really come to know—practically, through experience for yourselves—the love of Christ, which far surpasses mere knowledge (without experience): that you may be filled (through all your being) unto all the fullness of God—that is, may have the richest measure of the divine Presence, and become a body wholly filled and flooded with God Himself!
Ephesians 3:19 Amplified Bible

> *I will simply thank You
> and praise You ... always.*

Pineapple Weed
Matricaria matricarioides

A Fairbanks friend introduced me to Pineapple Weed. She pointed it out in her yard, called it Alaskan Chamomile, and praised the wonderful tea it made. The first name seems more applicable as the greenish-yellow flower heads look and smell like pineapples.

A pot of tea, shared with a friend, is one of life's purest pleasures. I am reminded of the little Girl Scout song:
*"Make new friends, but keep the old.
One is silver and the other is gold."*
Yes, friends are a treasure. How grateful I am for them, those I've known for half a century and those brand new.

As precious as my friends are to me, they can't compare to my Friend of all friends...JESUS!

    He is: The Bread of Life—satisfying that nameless hunger within me.

    The Living Water—quenching my thirst for the world around me.

    The Light of the World—filling me with the brightness of His presence.

    The Good Shepherd—nudging me back when I stray.

    The True Vine—supplying daily sustenance to this branch.

    The Beginning and the End—equipping me for all that's in between.

    The Resurrection and the Life—giving me life abundant now and life eternal later on.

Dear Father God, I can never comprehend the love and grace that created my friendship with Your Son. The idea that He values MY friendship as I value HIS, can never be understood by my mortal mind. I will simply thank You and praise You...always.

Henceforth I call you not servants; for the servant knoweth not what his lord doeth: but I have called you friends; for all things that I have heard of my Father I have made known unto you.
John 15:15

> *Things are not always what they seem to be.*

Chickweed
Stellaria media

It once would have taken a lot of grace to classify Chickweed as a wildflower. I much preferred to call it a weed—the weed I loved to hate! My dislike for Chickweed went back to my gardening days in Fairbanks. Being fragile, it invariably broke off above the ground when I tried to pull it out. Being strong and determined, the roots kept right on growing, creating new growth that eventually covered the entire garden—it was a yearly war—one I always lost!

We all carry memories from the past. Raw, festering memories, in various sizes, needing to be healed. The negative mental image that popped into my mind at the very mention of Chickweed needed the healing touch of the Great Physician.

Only as my hate was replaced by love and acceptance could I see the following positive lessons in this delicate wildflower:

1. Chickweed is edible; it may be cooked or eaten raw in salads. Spruce hens love the seeds. It has purpose and usefulness. God created us all for a purpose, even those we would judge as good-for-nothing.
2. Despite repeated attacks by humans, Chickweed determines to grow and spread. We need that kind of spiritual determination, no matter the attacks of Satan.
3. The tiny flowers of Chickweed look as though they have ten petals. Upon closer inspection, we find the ten petals are actually five deeply divided ones. Things are not always what they seem to be; with God, when it seems nothing's happening, something's happening!
4. No amount of mowing will eradicate Chickweed. It must be removed by the roots—a slow, tedious task. Hurts of the heart don't go away easily either. We must allow the Lord to go to the root of the pain to begin His healing process.

Great Physician of the Soul, negative memories, no matter their size, can rob us of the ability to be teachable and growing today, and to be unteachable is to live in spiritual stagnancy. Give us courage to come for healing, and faith to know you won't turn us away.

Ask, and it shall be given you; seek and you shall find; knock, and it shall be opened for you.

Matthew 7:7

We shall be like him; for we shall see him as he is.

Twin Flower
Linnaea borealis

The clear, pure water of the Fox Spring, north of Fairbanks, is a sharp contrast to the iron laden water of numerous local wells. We, and many others, went there often for drinking water. On the hillside above the spring reside delicate trailing plants called Twin Flowers, named for the two pinkish-white bell-shaped flowers on each stem.

Growing up as an only child, I longed for a sibling. Being a twin would have been absolutely wonderful. Gratefully, there are people in my life today whom I have unofficially adopted as brothers and sisters.

As appreciative as I am of my adopted relatives, they dim in value compared to my adoption into God's family. Imagine! Me—His child! Jesus—my Elder Brother! As Elder Brother, He looks out for me when I'm unaware of dangerous places. He cautions me when I start to develop attitudes that aren't good for me—or the family. And He stands in front of me when that bully named Satan attacks.

I want to "grow up and be just like Him." He knows that's tough to do right now, but He encourages me to press on because someday, in a moment, in the twinkling of an eye, it will happen!

Precious Father, Thank you that I have Jesus to pattern my life after. Help me to so live that when folks look at me, they'll see the family resemblance and think of Him.

Beloved, now are we the children of God, and it doth not yet appear what we shall be, but we know that, when he shall appear, we shall be like him; for we shall see him as he is.
I John 3:2

When Fireweed goes to seed, the summer's over.

Fireweed
Epilobium angustifolium ssp. angustifolium

Fireweed is one of Alaska's most striking and prolific wildflowers. A versatile plant, the leaves are edible, and the young stalks may be peeled and eaten like asparagus. The flowers make wonderful jelly or honey. Some folks believe the sweet perfume of its magnificent blooms will even heal battered emotions.

Alaskans have a saying, "When Fireweed goes to seed, the summer's over." By mid August the plants will indeed be dispelling white cottony seeds that fly about, reminiscent of snowflakes soon to come. Fireweed will grace our wildflower adventure from Fairbanks to Homer. Gratefully, we will still see enough blossoms to envision the glory we missed in July!

Fireweed's common name is derived from its ability to revegetate after a fire. It is able to do this because of deep roots that escape damage. It is Fireweed's ability to grow again in burned out places that provides a lesson. Many are the fires that would destroy us: flames of disappointment, flames of betrayal and bereavement, flames of financial loss. Nevertheless, if our spiritual roots go down deep into the soil of faith and commitment, we will grow and bloom again.

Compassionate Father, the strong essence of Your love is a sweet fragrance when trials blaze my way. By faith I face tomorrow and look forward to that day when You will greet my faithfulness with Your ultimate "Well done!"

These trials are only to test your faith, to see whether or not it is strong and pure. It is being tested as fire tests gold and purifies it—and your faith is far more precious to God than mere gold; so if your faith remains strong after being tried in the test tube of fiery trials, it will bring you much praise and glory and honor on the day of his return.
1 Peter 1:7 Living Bible

Use me in whatever way You desire.

Dwarf Dogwood
Cornus canadensis

The species of North American Dogwood vary from trees to small wildflowers. The beautiful little Dwarf Dogwood ranges the entire length of our Alaskan adventure. Friends who lived a little south of Nenana were blessed with it. The plants provided beautiful ground cover over a large portion of their yard.

In legend, the Dogwood tree was large like the oak. It was used to make the cross on which Jesus died. Afterward it never grew big enough to be used for such an awful deed. As a memorial of Christ's passion, its blossoms grew in the shape of a cross; its petals bore nail prints; and a crown of thorns graced the center of each of its flowers.

I remember the legend of the tree when I see the wildflower, and I remember the cross of Christ when I see the four white petal-like leaves arranged in cross fashion. The whorl of green leaves beneath remind me of those who stood around in ridicule or in sorrow as He died. The tiny flowers, representing the crown of thorns He wore, hold the promise of a cluster of bright red fall berries. We believers are the fruit of His suffering; blood bought sinners, saved by grace.

I scarce can take in the love of a Father who would send His only Son to die that awful death for me; I can hardly comprehend the love and willingness of a Son who said, "Yes, Father, I will go and I will die for her sins, for the sins of the whole world.

Dear Saviour, in taking my place on Calvary You have given me forgiveness for my sins, acceptance into Your family and the promise of eternal life. Simple lip service will never express my gratitude. I do no less than relinquish my life—all my hopes, dreams and plans—to Your will. Use me in whatever way You desire.

As for me, God forbid that I should boast about anything except the cross of our Lord Jesus Christ. Because of that cross my interest in all the attractive things of the world was killed long ago, and the world's interest in me is also long dead.

Galatians 6:14 Living Bible

How precious You are to me as Reprover, Guide, Teacher and Comforter.

Arctic Lupine
Lupinus arcticus

Lupine, one of Alaska's most beautiful wildflowers, love to locate along roadsides. The ones seen growing along the Clear turn-off provide our spiritual lesson.

Clear Air Force Base was established back in 1960 as part of the Ballistic Missile Early Warning System. The Lupines grow in sight of, perhaps in the shadow, of enormous radar antennas, once used to provide early warning of an approaching enemy attack.

We Christians need spiritual radar. We need access to an early warning system to live lives that glorify Christ. Radar is needed to warn us when our priorities get mixed up, when our thinking is self-centered and our leaning is away from His will. We need something to alert us to those fiery darts Satan constantly throws at us.

Within us, we have such a system, in the Person of the Holy Spirit. He desires free reign of our hearts, complete command of all the controls. Then He will be able to quickly make known to us thoughts, attitudes and actions that if not dealt with, will "blow apart" our joy within and witness without.

Dear Holy Spirit, to be Your dwelling place—Oh, what grace without measure! Welcome! How I need Your presence within me. How precious You are to me as Reprover, Guide, Teacher and Comforter.

He who dwells in the secret place of the most High shall remain stable and fixed under the shadow of the Almighty (whose power no foe can withstand). I will say of the Lord, He is my refuge and my fortress, my God, on Him I lean and rely, and in Him I confidently trust!
Psalms 91:1, 2 Amplified Bible

> I hold tightly to You and
> Your eternal promise of spring.

Pasque Flower
Pulsatilla patens

Six miles south of the Clear turn-off lay the remains of the Liaho Trailer Park. It was there, back in 1961, in an 8-by 45-foot trailer, we spent our first Alaska winter. We misplaced Southerners shivered through temperatures of -70 degrees, days of 21 hours of darkness, and snow by the foot!

What a welcomed sight lovely Pasque Flower was the following spring. The large, purple cup-shaped flowers seemed eager to appear. Whenever the snow would melt in a sunny area, a flower would pop up as if to say, "Cheer up! It's Spring! Alaska is not all cold, dark and snow covered. I bring you the promise of 80 degree days, 20+ hours of sunshine, baseball games, picnics and fishing under a midnight sun."

Our God perfectly planned the seasons. His promise to Noah is still true today. "While the earth remaineth, seedtime and harvest, and cold and heat, and summer and winter, and day and night shall not cease. (Genesis 8:22) Is He not God of the seasons of the soul as well? As we endure a spiritual winter, we can be sure spring will follow. Spring—with its sunny days of renewed hope, stronger faith and deeper commitment.

Faithful God, sometimes my heart feels the cold grayness of winter. During such times I hold tightly to You and Your eternal promise of spring. I rejoice that no matter the harshness of winter, spring will burst forth with new reasons to praise and glorify You.

After you have suffered awhile, our God, who is full of kindness through Christ, will give you his eternal glory. He personally will come and pick you up and set you firmly in place, and make you stronger then ever.

1 Peter 5:10 Living Bible

> I need Your ... holy resolve to keep keeping on—regardless of the winds!

Siberian Aster
Aster sibiricus

Continuing our tour south, we'll stop a few miles past Healy. My husband and I stopped there two summers ago to change drivers and I found a little Siberian Aster. If this area of the Parks Highway could enter a contest to become "Wind Capital of the World," it would probably succeed. The wind seems to be always blowing in Healy—and blowing hard.

Growing along the highway, the little Aster leaned under the pressure of the wind. I wondered if it was ever able to stand up straight. Undoubtedly the constant pressure of the wind was frustrating and exhausting. "Determination!" I thought, "That little plant is determined to withstand the harshness of its environment and bloom." Surely the secret of its endurance is roots that grow deep and cling tightly to the soil and rocks below the surface of the ground.

Many of us today feel the constant pressure of the wind in our lives: occupational winds, family winds, financial winds, winds of social involvement and church responsibilities. We need determination too! The incessant pressure of finding balance will leave us stooped and bent unless the spiritual roots of our lives go down deep into the soil of faith in God and commitment to His will.

Father God, standing straight is often hard, Daily demands push and pull at me constantly. I need Your strength and wisdom to stand upright in a world that pressures me to lean away from You; and holy resolve to keep keeping on—regardless of the winds!

We are pressed on every side by troubles, but not crushed and broken. We are perplexed because we don't know why things happen as they do, but we don't give up and quit.
2 Corinthians 4:8 Living Bible

I must admit that most of what obstructs my view of You is of my own making.

Watermelon Berry
Streptopus amplexifolius

The many seeded, bright red oval fruits give Watermelon Berry its name. In the fall they may be gathered and made into jelly or syrup.

Watermelon Berry and I met about 50 miles south of the entrance to Denali Park. It was growing in a new rest area complete with picnic tables, a nature walking trail, and telescopes for viewing "The Mountain."

Mt. McKinley, the highest mountain in North America, is called Denali by the native people. Denali means Great One. Watermelon Berry chose to grow, where on a clear day, Denali—Great One—can be seen in all its glory.

Let us follow Watermelon Berry's example; position ourselves in full view of our great God. Some days it may be hard to see Him. The view may be blocked by clouds beyond our control: sickness, accident, financial disaster, or loss of some kind. On those days, it is by faith, not by sight that we know He is there. There are other days when we block the view ourselves, with self-made clouds of unforgiveness, anger, doubt, unlovely attitudes or everyday "busyness." It is then that the great wind of repentance, of saying, "I'm sorry, forgive me," will blow the clouds away, revealing again His beauty and majesty.

Patient Father God, if I am honest I must admit that most of what obstructs my view of You is of my own making. Forgive me. How swiftly the sweet fellowship we share can cloud over. May the Holy Spirit within me be a spiritual barometer, alerting me to any danger that threatens to block my view of You.

Looking unto Jesus, the author and finisher of our faith, who for the joy that was set before him endured the cross, despising the shame, and is set down at the right hand of the throne of God.
Hebrews 12:2

I have always been wrapped securely in Your eternal love.

Arctic Daisy
Chrysanthemum arcticum ssp. arctzcum

South of Denali Park, we enter Arctic Daisy country. The remainder of our trip to Homer will be sprinkled with this lovely wildflower that is actually a member of the chrysanthemum family.

Long ago, when we put away our dolls and tea sets, stopped playing hopscotch, jump rope and jacks, and boys suddenly changed from dumb to neat, my friends and I played a game using daisies. Remember? It was called, "He Loves Me, He Loves Me Not." Daises became a mystical avenue by which we could determine if a "certain" boy returned our affections. Many a lovely daisy fell victim as we slowly plucked the petals, one by one, and chanted the little jingle. We were all smiles and giggles when the last petal was a "He loves me," and devastated when it was "He loves me not."

My joy and gratitude are boundless when I remember that I don't have to play the "He Loves Me, He Loves Me Not" game to determine the affections of God. I have put away the childish notion that if I'm a good person, wife, mother, neighbor or church member, He will love me. But if I'm bad and don't live up to His expectations nor mine, He won't. He is my Heavenly Father. I can't earn His love; He gives it freely and unconditionally.

Loving Father, on occasion, You've had to use Your spiritual paddle when I have been a stubborn, willful, disobedient child. Throughout the chastening, You never disowned me. I have always been wrapped securely in Your eternal love. I want to serve You and follow Your precepts--not SO that You will love me, but because You DO!

And may you be able to feel and understand, as all God's children should, how long, how wide, how deep, and how high his love really is; and to experience this love for yourselves, though it is so great that you will never see the end of it or fully know or understand it. And so at last you will be filled up with God himself.

Ephesians 3:18, 19 Living Bible

Create in me a deep hunger for the Bread of Life,... and the fruit of the Holy Spirit.

Butter and Eggs
Linaria vulgaris

Although Butter and Eggs grew in my yard in Fairbanks, I've never seen it in such abundance as I did surrounding the train station in Wasilla, August 1995. So prolific were these splendid flowers that the station and tracks seem to sit on a variegated carpet of yellow, orange and green.

The name aptly describes the color of these snapdragon-like wildflowers. The blossoms are the yellow color of butter with a palate area the darker yellow-orange of an egg yolk.

Butter and Eggs start me to thinking of breakfast; add some toast and jelly, a cup of coffee or tea, and you have it! Rarely do I miss this morning meal. I believe in its nutritional advantages. On a particularly busy day, I'll even get up earlier to have time to eat. That's how dedicated I am to breakfast!

Breakfast for my soul, however, is pretty haphazard, sometimes yes, sometimes no. I am often too busy, running late, or simply more desirous to get on to other things. Could there be a correlation between the days I have no time for early morning spiritual food—prayer and the Word—and the days I "fall apart" by noon, having little patience, kindness or compassion for the people in my world? Would I be more apt to see positives and exhibit joy and smiles if I made the time to feed my soul at the beginning of each day?

Dear Father, forgive me. Sometimes I get my priorities all mixed up. I forget "first things first." I am more interested in dining at my table than at Yours. Create in me a deep hunger for the Bread of Life, the milk of Your Word and the fruit of the Holy Spirit.

My voice shalt thou hear in the morning, O Lord; in the morning will I direct my prayer unto thee, and will look up.
Psalm 5:3

*Give me grace and wisdom
to keep the scale level!*

Monkshood
Aconitum delphinifolium ssp. delphinifolium

True to its name, the bluish-purple flowers of Monkshood, scattered on a long stem above the leaves, do look like little round heads covered with hoods.

Upon seeing Monkshood for the first time, my imagination whisked me away to a lovely old stone monastery on a wooded hillside in some far off place. The brown-robed monks, with their hoods pulled low, walked about with hands folded in prayer, or sat for undisturbed hours reading God's Holy Word, or knelt before the altar for long periods of simple praise.

"That's for me!" I exclaimed excitedly. To spend my days in sweet fellowship with God, away from all the demands and frustrations of my life seemed so inviting!

I was ready to start packing! Then I remembered there was laundry to do, the house to clean, dinner to cook, dishes to wash, a class to teach, lonely nursing home folks to visit, a sick neighbor to help...the list went on and on.

"Balance." the Holy Spirit whispered. "You need balance. The weight of everyday responsibilities should balance your time spent with the Father, but never outweigh it. Nor should your time with the Father be so all consuming that you have no time left to care for those He's put in your keeping or called you to minister to."

I find the hardest part of staying balanced is knowing when to say NO! Even good things can throw me off balance if they diminish sacred time with my Lord, or my duties to those in my life.

Dear Father God, living a balanced life honors You. It also brings peace, joy and contentment. Give me grace and wisdom to keep the scale level!

Let me be weighed in an even balance, that God may know mine integrity.

Job 31:6

Please control my tongue that I may never abuse your precious gift of language.

Rattlebox
Rhinanthus minor

Wasilla is a good place to look for Rattlebox. By August, the small inconspicuous flowers are rather old, faded or have fallen away altogether. However, our late summer arrival is an ideal time to see and experience how this wildflower got its name. The slightly hairy calyx has formed an urn-like structure that actually does rattle when the seeds are mature.

I am reminded of an expression I've always heard to describe someone who talks nonstop on a particular subject. We say they "rattle on."

Some folks "rattle on" about how they're being mistreated; how unjust is the hand life has dealt them. Then there are the folks who "rattle on" about the failings of everyone else, leaving no time to look at their own. Saddest of all, perhaps, are the folks who "rattle on" just to be talking. They seem to be more interested in the quantity rather than the quality of their words.

"Rattling on" comes natural to us mortals. Left unchecked, the tongue in each of us will find a subject to "rattle on" about. The tongue can so easily inflict pain on another. The old adage, "Sticks and stones may break my bones, but words can never hurt me," IS A LIE! It takes determination and purpose to be certain that our speech is filled with words of encouragement, inspiration, concern, and love.

Holy Father, please control my tongue that I may never abuse your precious gift of language. When I'm tempted to "rattle on" about another, help me instead, to rattle on about Your matchless grace to me!

Let the words of my mouth, and the meditations of my heart, be acceptable in thy sight, O Lord, my strength, and my redeemer.
Psalms 19:14

> *I . . . joyfully exclaim, "Thank You, thank You, thank You!"*

Grass of Parnassus
Parnassia palustris

Grass of Parnassus ranges over most of Alaska, except the Aleutian Chain. The name is a misnomer. Parnassus is a mountain in Greece where this wildflower does NOT grow and in no way does it resemble grass.

The leathery, almost heart shaped leaves grow at the base of the plant, with one smaller leaf clasping the stem about half way between the ground and the flower. The five petaled flower, perhaps an inch across, is white with green veins. The central seedpod is very prominent.

My description doesn't sound too exciting and any attempt to illustrate Grass of Parnassus would disappoint even the greatest of artists. But when I pick a flower and hold it close, my lips cannot keep silent. Some exclamation of wonder must burst forth! The absolute beauty of this symmetrical wildflower is like joining with the Psalmist and declaring, "Be still, and know that I am God!" It is like looking into the face of our creator God and KNOWING that HE IS!

Our choices of flowers would probably be different for each of us, but as lovers of wildflowers, we can all relate to the awe I find in Grass of Parnassus. Of all the wildflowers I have experienced, it is undoubtedly Grass of Parnassus that beckons me to "take off my shoes, for I am standing on holy ground!"

Creator God, although I have stood upon tall mountains, looked up at towering trees, watched wheat fields dance with the wind, listened to ocean waves crashing on the beach, it is in the awesome beauty and diversity of wildflowers that I most clearly see Your unconditional love. I am blessed to be one of the few who stop, notice and joyfully exclaim, "Thank You, thank You, thank You!"

Oh, come, let us worship and bow down; let us kneel before the Lord our maker.

Psalm 95:6

> Fear not, then; you are of more value than many sparrows.

Ladies Tresses
Spiranthes Romanzoffiana cham.

The little wildflower called Ladies Tresses belongs to the orchid family. Tiny three petaled, aromatic, white flowers spiral around the stem in rows reminiscent of my granddaughter's blond curls or powdered wigs and carefully coiffured ringlets of hair about ladies heads in time past.

When Jesus talked of the cost and compensations of discipleship, He commented on the hair of our head, sandwiching His comment between two verses about sparrows. Sparrows are so valuable that He notes the fall of even one, yet a flock is of lesser value than we, His children.

Discipleship may be costly and often scary but we can rest in the assurance that God is able to care for us.

"As compassionate, caring and watchful as I am of sparrows," our Lord declares, "Let me tell you of the extent I watch over you! I keep so close a check on you that I have counted the number of hairs on your head! Would I bother if you were not of indescribable worth to me?"

Whatever the cost of following Christ, it is little compared to the satisfaction of knowing about His detailed awareness of me.

Concerned Father, to know You care so intimately for me turns fear into courage, obedience into privilege, and service into joyful expectation.

Are not two little sparrows sold for a penny? And yet not one of them will fall to the ground without your Father's leave and notice. But even the very hairs of your head are all numbered. Fear not, then; you are of more value than many sparrows.
Matthew 10:29-31

> *Arise, my love, my fair one, and come away.*

Mountain Harebell
Campanula lasiocarpa

If you follow the Hatcher Pass road east out of Wasilla you will wind your way along the beautiful Susitna River, up past the tree line of cottonwood and willow, to the old Independence Gold Mine, now a state park. My daughter, granddaughters, and I were hiking there on alpine slopes the day I found the Mountain Harebell.

It is such a fragile plant. The large upright bell-shaped flower sits on a stem no more than four inches tall. Mountain Harebell reminds me of the gentle call from Song of Solomon, chapter 2, verses 10-13: "My beloved spoke, and said unto me, Rise up, my love, my fair one, and come away. For, lo, the winter is past, the rain is over and gone. The flowers appear on the earth; the time of the singing of birds has come, and the voice of the turtledove is heard in our land. The fig tree putteth forth her green figs, and the vines with the tender grapes give forth fragrance. Arise, my love, my fair one, and come away."

No greater example of the need to "come away" can we find than in our Lord Himself. Wisdom for decision-making, strength for service, and just plain weariness demanded He "come away" often. So let us too, climb above the tree line of our everyday lives, above the demands and pressures of everyday living. "Come away" to quiet, undisturbed fellowship with God. This special kind of fellowship can be experienced only as we climb above the "busyness" of life.

Wise, gentle Jesus, it was never convenient for you to be alone; the sick in body and spirit were always about, needing Your help. Yet you knew it was solitude with the Father that energized, empowered and enabled You to get through the day. Therefore, You MADE time to "come away." Give me holy wisdom to know and to do likewise.

And straight way he constrained his disciples to get into the boat, and to go to the other side before him unto Bethsaida while he sent the people away. And when he had sent them away, he departed into a mountain to pray.

Mark 6:45, 46

> *I desire to "wear out" not "rust out" in my service to You.*

False Hellebore
Veratrum viride Ait.

Another inhabiter of the slopes of Hatcher Pass is a tall, but not impressive, plant called False Hellebore. The large linear leaves clasp the stem below a flower stalk covered with greenish blooms. I was even less impressed when I learned it is very poisonous. It has little in common with the true Hellebores that originated in Europe and Asia. Perhaps that is the reason it is called false.

Paul, the great apostle of the Christian faith, was a man of infinite courage. He endured shipwreck, beatings, stoning and jail. Through them all, he was never afraid to speak for his Lord. However, Paul did have one fear. He feared being labeled false. Having preached to others, he feared he might fall under the poisonous effects of lack of self-discipline, complacency and spiritual neglect. To be seen as unfit, to be set-aside in the race of life, was the one great dread of Paul's life.

We share Paul's concern. To finish our race well, we must be on guard. In little ways, the enemy of our soul would delight in tripping us up. Changes that would disqualify us, from a fruitful walk, come on without warning. Perhaps we become so entangled with "good" things of the world that we have no time left for the spiritual. The aches and pains of growing older can become excuses to lessen our commitment. Or God does not answer our prayers in a certain situation as we think He should and we are hurt, angry and withdraw. Sometimes we allow the failings and hypocrisy in the lives of others to justify our pulling back and quitting.

Faithful God, I desire to "wear out" not "rust out" in my service to You. I, like Paul, fear to be set aside. In this great race of life, help me to be aware when I deviate, even the slightest, from the course you have marked out for me.

Like an athlete I punish my body, treating it roughly, training it to do what it should, not what it wants to. Otherwise I fear that after enlisting others for the race, I myself might be declared unfit and ordered to stand aside.
1 Corinthians 9:27 Living Bible

Places to relocate are everywhere if we will only listen to God's call.

Blue Flag
Iris setosa

The large showy flowers of Blue Flags put on a magnificent display of color and design. The marshlands between Wasilla and Anchorage are a perfect location for the show!

Many wildflowers are particular where they grow, but not Blue Flags. Although they prefer boggy places, Blue Flags will not refuse to grow in hard ground away from contact with surface water. They also transplant easily. Recent transplants grow happily in my daughter's yard and filled my flowerbeds with beauty for many years.

Blue Flags are not self-centered in their philosophy of life. "I like where I am," they say. "I am comfortable here, but go ahead and move me where you want me; wherever I can bring the most joy to you."

The Apostle Paul shared that philosophy. He never allowed himself to "settle-in" to one place and be tied down by "things" and "stuff." When the Holy Spirit said, "Moving time," Paul was ready.

The Holy Spirit's call to relocate comes to us too: perhaps from soaking in the riches of His Word in our Adult Sunday School Class to changing soaky diapers in the church nursery; from relaxing in our recliner to "relaxing" with a group of teenagers on an outing; or transplanting time required to pursue a favorite hobby or sport into time in a nursing home pursuing and encouraging lonely hearts. Places to relocate are everywhere if we will only listen to God's call.

Heavenly Father, in the light of eternity, help me to see that MY "things" I hold so dear and MY "stuff" that seems so important are really of little value. Being in and doing God's will is what counts.

I have strength for all things in Christ Who empowers me—I am ready for anything and equal to anything through Him Who infuses inner strength into me, [that is, I am self-sufficient in Christ's sufficiency.]
Philippians 4:13 Amplified Bible

Our prayers seem to only hover about as breath on a cold day.

Alaska Cotton
Eriophorum Scheuchzeri

Rubber boots are almost a necessity if one wants to inspect Alaska Cotton up close. It grows from Fairbanks to Homer, but most always in wet, peaty soil, right at the edge of lakes or shallow water, or water-filled ditches. The flowers are inconspicuous and top a stalk that later develops into a fluffy white seed head.

I imagine Alaska Cotton acquiring its name from some miner as he plucked a wild goose and noted the similarity between its fluffy down and the plant growing nearby. He wondered at the possibility of using the seed heads to insulate his vest, jacket or cap. Being a Southerner by birth, in the far north searching for gold, the white seed heads reminded him of cotton fields back home so he called the plant cotton, Alaska Cotton.

My story of how Alaska Cotton got its name is only true in that the plant has been used to insulate clothing. Thinking of insulation, I am reminded of what I call "cold soul days." We all experience them now and then. Sometimes we can pinpoint the reason, sometimes we can't. God seems distant. We feel like we're in the North Pole and He is in Florida. Our prayers seem to only hover about as breath on a cold day.

God's Word, through the pen of the prophet Isaiah, has the answer for our "cold soul days." The solution is to add a new garment to our spiritual wardrobe. Guaranteed to warm the heart, soul, and spirit, this new attire cost no money, requires neither talent nor training to put on. One size fits all. Isaiah calls it "the garment of praise."

Loving Father, continuing in a "cold soul day" is by choice, not necessity. Praise Your precious name! Whenever I feel the temperature dropping, I need only fill my mind and lips with praise to feel my heart warm up to Your amazing love and grace.

To appoint unto those who mourn in Zion, to give unto them beauty for ashes, the oil of joy for mourning, the garment of praise for the spirit of heaviness, that they may be called trees of righteousness, the planting of the Lord, that He might be glorified.

Isaiah 61:3

> *For all eternity I shall praise You for Your AMAZING GRACE.*

Blue Columbine
Aquilegia brevistylla

Ninilchik, with its Russian heritage, is a busy fishing mecca twenty-five miles north of Homer. We were vacationing in the state campground, hoping for a record breaking halibut catch. My first visit to the outhouse held an unexpected surprise. Seemingly growing out of a crack in the wall grew a Blue Columbine.

I stood transfixed; my eyes beheld its awe-inspiring beauty until my nose encouraged me to tend to business quickly and get back to our campsite. Each time I visited the outhouse I was both captivated with the loveliness of the Blue Columbine and repulsed by the smell of it's surroundings.

Looking back I see a beautiful picture of Grace. We mortals tend to think of ourselves as basically good. We all know folks, even those who make no profession of faith, whom we call "good people." What a different story the Bible tells. The prophet Isaiah said, "we are as an unclean thing, our righteousness is as filthy rags." Several centuries later, Paul agreed when he wrote, "for I know that in me dwelleth no good thing."

God sent His Holy Son to dwell among us, not from a distance where He could not smell the stench of our sinfulness, but in the midst of us. He is the "Lily of the Valley" and the "Rose of Sharon;" BEAUTY amid ugliness, PERFECTION among the flawed and broken, HOLINESS among the defiled and corrupt. And then on the cross of Calvary, He died for us. In love He removed, for all time, our sins with their putrid smell and replaced them with the sweet fragrance of His forgiveness.

Father-God, my finite mind can never truly understand Your gift of unmerited favor. It is enough that You extended it and I accepted it. For all eternity I shall praise You for Your AMAZING GRACE.

For all have sinned and come short of the glory of God. For the wages of sin is death, but the gift of God is eternal life through Jesus Christ our Lord.

Romans 3:23, 6:23

> May your roots go down deep into
> the soil of God's marvelous love.

Beach Fleabane
Senecio pseudo-arnica

The name of this daisy-like wildflower discloses much about the plant. It's related to all Fleabanes, named for the belief that hanging them in the home would ward off fleas. And from its name we know where to find it—at home on the beach.

Beach Fleabane and I share a love of the seaside; the sparkle of the water, the constant mighty sound of the waves rolling upon the shore, the happy squawk of sea birds flying about, the wonder of anemones and jelly fish left in tidal pools, the sand just right for castles and frog houses. Yes, I could be right at home on the beach too.

Revelation 3:20 pictures Jesus knocking on the door of the heart. He desires not only to dwell there, but to be at home. I understand that He doesn't want to simply sit around, treated like a guest to be pampered and waited on. No, for Him to be comfortable and really feel at home in my heart, He wants to share the chores. Together we:

Hang the walls with His Word to ward off the devil,

Brush down the cobwebs of the past,

Dust away lingering resentments and anger,

Sweep out unforgiveness,

Polish tarnished commitment,

Mop up tears of broken dreams and unmet goals,

Clean out words that put-down, hurt or ridicule,

Replace them with the soft language of love and understanding.

Dear Jesus, I could never keep the home of my heart clean without Your help. My daily "to do" list needs Your check and approval too. Please prioritize it according to Your will.

And I pray that Christ will be more and more at home in Your hearts, living within you as you trust in Him. May your roots go down deep into the soil of God's marvelous love.

Ephesians 4:17 Living Bible

Loving Lord, may this truth deflate the pride carried by some Fireweeds.

Willow Herb
Epilobium species

Bishop's Beach in Homer, with it's floatable wildlife trail, is home to many wildflowers. The Willow Herb grows there as well as in the hills above. Plants I have found average less than a foot in height.

I was sure I had never seen this spindly, rather fragile wildflower before, yet something about it seemed familiar. Then I noticed the seedpods. They were identical to Fireweed! Could they belong to the same family? My hunch was right; the little Willow Herb does indeed belong to the same family of wildflowers as the tall and prolific Fireweed. They share identically shaped seedpods, even to the fluffy seeds carried away by the wind.

Everybody knows Fireweed! It graces whole hillsides, grows for miles along highways, can even be seen from airplanes. In contrast, Willow Herb is small and inconspicuous—one has to practically stumble over it to notice it. Although it is found over the state, as is Fireweed, few people ever see it.

These wildflowers, members of the same family, remind me of a great truth. Fireweeds are the Christians with talent and calling we all see: pastors, teachers, and the musically gifted. They are continually in view, praised and thanked for their service and talents. Willow Herbs are Christians who are faithful without recognition, who labor where folks don't see. In my church, the Willow Herbs are those who vacuum the carpet, prepare the communion table, straighten the basement before and after a church supper, print the bulletin every week and are present on every mowing day. The service they do tend to be overlooked and taken for granted.

Although our callings, talents and gifts are varied and different, they are equal in Christ. All service performed out of love for Him, and the family, receive the same recognition in His sight.

Loving Lord, may this truth deflate the pride carried by some Fireweeds. May it joyously inflate the feelings of those Willow Herbs who see their service as "less than" and unimportant.

Behold, how good and how pleasant it is for brethren to dwell together in unity.

Psalm 133:1

> *Coat our joyousness with infectious love as we "spoon" it out to a needy world.*

Yellow Monkeyflower
Mimulus guttatus

Named for the grinning look of the blossoms, Monkeyflowers evoke smiles and laughter within me. In memory I am a little girl again, watching the playful monkeys romp about Monkey Island in the San Francisco Zoo. If God had in mind to create a creature simply to bring laughter to little children, He succeeded in the monkey.

So often we take ourselves too seriously. We tend to gravitate toward sadness and "poor me" mindsets. We need someone to help us see the humor in life. I think of people who have touched my life with laughter, wonderful friends and acquaintances with "monkey personalities." They enter a room and it's brighter. They enter a conversation and soon it's filled with smiles and mirth.

Wise King Solomon compares a merry heart to a dose of medicine. Medicine that helps:

Lower stress pressure,
Cool the temperature of a hot temper,
Relieve the ache inflicted by wayward children,
Soothe the feelings hurt by others,
Mend the heart broken by loss,
Lighten a discouraged spirit.

Another man of wisdom, Abraham Lincoln, said, "Most folks are about as happy as they make up their minds to be." Although happiness is a choice each of us must make for ourselves, having others rooting for us in the grandstands helps make our decision easier.

Compassionate Jesus, lots of folks need their lives brightened and their loads lightened with a dose of laughter. Few of us are blessed with "monkey personalities", but we all have JOY in Christ. Coat our joyousness with infectious love as we "spoon" it out to a needy world.

A merry heart doeth good like a medicine, but a broken spirit drieth the bones

Proverbs 17:22

The little we know of Heaven is nothing compared to the reality that awaits us.

Goldenrod
Solidago lepida

Several species of Goldenrod reside in Alaska. They are not the hay fever culprits as is commonly believed; their pollen is sticky and rarely airborne. So forget the negative reaction Goldenrod may bring to mind and rejoice instead in the positive application we'll see in this lovely wildflower.

With eyes of faith I envision home when I see Goldenrod. Not my former home in majestic Alaska, nor my present home among the beautiful hills and creeks of northwest Alabama. Those were, and are, temporary. It is a future home I see. The Carpenter of Nazareth has gone on to begin construction. I haven't seen the blueprints, but He called it a mansion. It's located in a city of pure gold—even the streets! The wall around the city is supported by a foundation garnished with all kinds of precious stones. The gates are made of pearls. The home the Carpenter is building must be some kind of fancy place to be allowed in a city like that!

Yes, Goldenrod gets me thinking of Heaven, a prepared place for a prepared people. Can you believe I don't have to even pack for this move? In preparation, the Carpenter said all I needed was a ticket. I gladly offered to buy it, but He said, "No, it can't be earned or bought. It's free!" He removed it from a book called Amazing Grace, stamped it "PAID IN FULL ON CALVARY", and GAVE it to me!

Gracious Father-God, I suspect the little we know of Heaven is nothing compared to the reality that awaits us. I pray the Holy Spirit will fill my life and help me use the time before departure to enlist others to come along with me.

Eye hath not seen, nor ear heard, neither have entered into the heart of man the things which God hath prepared for them that love him.
1 Corinthians 2:9

Hidden inside, the flowers of my heart were wilted by guilt, confusion and pain.

Yellow Paintbrush
Castilleja scrophulariaceae

The flowers of all Paintbrushes are minute. They are hidden by the bracts (modified leaves) surrounding them. Homer is home to many Yellow Paintbrushes. Mistakenly we think we are looking at a lovely meadow of yellow flowers when actually we are admiring a meadow of leaves.

I was once like the Yellow Paintbrush. Hidden inside, the flowers of my heart were wilted by guilt, confusion and pain. However, my outside was covered with "I'm fine, thank you," and a big smile. I allowed others to mistake my outside for my inside. I never gathered the courage to correct the wrong impression I had given.

Physically, emotionally and spiritually suffering paintbrushes need a fresh coating of the love of 1 Corinthians, chapter 13.

> Love is patient and kind, never jealous or envious, never boastful or proud, never haughty or selfish or rude. Love does not demand its own way. It is not irritable or touchy. It does not hold grudges and hardly notices when others do it wrong. It is never glad about injustice, but rejoices whenever truth wins out. If you love someone you will be loyal to him no matter what the cost. You will always believe in him, always expect the best of him, and always stand your ground in defending him.

Dear Holy Spirit, let me see others through your eyes that I may recognize the hurting folks in my world. Then direct me in ways of spreading a coat of Your great love over them.

There are three things that remain—faith, hope and love—and the greatest of these is love.

1 Corinthians 13:13 Living Bible

> *I choose to place
> the future in His hands.*

Pink Pyrola
Pyrola asarifolia

The Wynn Nature Center sets in the hills high above Homer. Stand with me there in the loveliest of meadows. Slowly breathe in the beauty of Kachemak Bay and the glistening glaciers and snow-capped peaks of the Kenai Mountains. Stately spruces dot the edges of the meadow. A myriad of wildflowers dwell in this pristine location. Among them are the little Pink Pyrolas.

Pyrolas are evergreens. Five petaled, nodding pink flowers grace 10-inch stems in summer. The rest of the year their basal rosette of shinning green leaves make lovely ground cover. Such a dwelling place as this, in the midst of such peace and beauty, must fill the roots of Pink Pyrola with gratitude and thanksgiving.

For many years, my dwelling place was in the past. Guilt, regrets, "should haves" and "wish I'd dones" kept me chained when I wanted to let go. I have also dwelt in the future; happiness, peace, and contentment waiting somewhere "out there" as soon as the people in my life met my expectations, or when I reached certain goals, or when I acquired certain possessions.

Enough of the past and the future! I choose to dwell in the Lord TODAY! I choose to allow Him to heal the past. I choose to place the future in His hands, having learned that He is the One who brings joyous satisfaction, not people, accomplishments or accumulations.

Dear Father-God, I need look no further when You are my abiding place. With thanksgiving, I'll just settle in, surrounded by the beauty of Your Holy Presence.

O Lord, thou hast been our dwelling place in all generations.
Psalm 90:1

> May my hunger for the meat
> of Your Word never be satisfied.

Forget-Me-Not
Myosotis alpestris, ssp. asiatica

Few people, having visited the great land of Alaska will ever forget its uniqueness, beauty and grandeur. Few people, having experienced the great love and forgiveness of God will ever forget Him either.

Many Christians, however, do forget the next step in faith. The dainty Forget-Me-Not is a reminder of this spiritual principle. Like several species of wildflowers, the tiny buds begin as a lovely pink and then turn a brilliant sky blue as they mature. Those pink and blue colors always remind me of babies. And what do babies do? They grow up!

And what do many Christians do? They forget to grow, to mature. They never develop an interest in being more than spiritual babies. They go through life nourished only by a few drops of milk on Sunday morning. They never try to feed themselves and thus never experience the deeper truths of God's Word. They are satisfied with a spiritual diet of milk, when the Lord would have them eat solid food.

The spiritual understanding of these perpetual babes in Christ never matures enough for them to see the difference between knowing Jesus as Savior and knowing Him as Lord. Sadly, they crawl through life, not ever knowing the pleasure of walking in sweet fellowship with Him.

Precious Lord, May my hunger for the meat of Your Word never be satisfied. May my desire to know You intimately result in a walk that hops, skips and runs joyously through life.

You have been Christians a long time now, and you ought to be teaching others, but instead you have dropped back to the place where you need someone to teach you all over again the very first principles of God's Word. You are like babies who can drink only milk, not old enough for solid food.
Hebrews: 5:12 Living Bible

That old nature will have little to do but sit and wave a flag of surrender!

Nettle
Urtica gracilis, U. lyallii

A burning sensation to the skin quickly notifies anyone who accidentally comes in contact with Nettle plants. Both leaf and stem are covered with soft stinging hairs that cause burning and develop into a rash. However, you don't have to look far for treatment. The plant itself is both the cause and the cure. Simply put on gloves, squeeze the juice from the stems and apply to the rash.

What a picture of we who bear the name of Jesus Christ. The Nettle plants represent us. At times we are the "Juice," eager to build up, encourage, help, eagerly sacrificing ourselves for others and the kingdom. At other times we are "leaves and stems," quick to hurt, knock down, ridicule, selfishly refusing to give time and service to the cause of Christ.

Paul understood the strife of the two natures within us. He not only understood, he struggled as well. He laments his predicament in Romans, chapter seven. We identify with his frustration, "When I want to do good, I don't, when I try not to do wrong, I do it anyway." We feel the cry of his heart, "O wretched man that I am, who shall deliver me from the body of this death?" We joyfully shout with him, "Thank God! It has been done by Jesus Christ our Lord. " He has set me free." Not completely free of the old nature, but free of the condemnation it brings. We'll loose a battle once in awhile, but the war has been won!

Precious Lord Jesus, as I allow the Holy Spirit to powerfully fill all the "nooks and crannies" of my heart, that old nature will have little to do but sit and wave a flag of surrender! Praise be Your holy name.

There is, therefore, now no condemnation to them who are in Christ Jesus. Those who let themselves be controlled by their lower natures live only to please themselves, but those who follow after the Holy Spirit find themselves doing those things that please God.
Romans 8:1,5 Living Bible

> *I never set my expectations above another's limitations.*

Wild Geranium
Geranium erianthum

We end our wildflower field trip in my daughter's yard, high in the hills above Homer.

The wild Geranium is also known as Cranesbill; named for the shape of the seedpod. These interesting pods are so constructed that as they dry, they build up tension until they burst at the seams. The five sections spring upward and can catapult their seeds as far as 22 feet.

In the Wild Geranium I am reminded to be thankful, never take for granted, the precious institution planned by God—the family! When I consider "family" I list:
1. My family of origin—those who came and went before me, who are part of who I am today.
2. My immediate family: my husband, children, grandchildren, parents, aunts, uncles, and cousins.
3. The family of faith I belong to, with whom I worship each week.

As blood relations, my secular family is related by birth. My spiritual family is connected by blood and birth also; the precious blood of Jesus shed for our sins, resulting in spiritual birth. Although the blood ties of both families make them strong, interaction between their members often is fragile.

Each family needs nurturing. They need big doses of acceptance, encouragement, love and time. Two way needs. As I give, so am I given. As I give time: telephone time, letter time, in person time— the blessing I give benefits me as well.

Yes, families are important. So each week I put aside all else and meet with my family of faith. And each year, I step aboard a jet plane and catapult myself 4,000 miles to spend time with my family in the Far North.

Father God, a member of your family, and mine, once said, "I never set my expectations above another's limitations." How wise, Lord. Give us grace to meet precious members of the family where they are and love them as we find them.

And let us consider one another in order to stir up love and good works, not forsaking the assembling of ourselves together, as is the manner of some but exhorting one another, and so much the more as you see the day approaching.   *Hebrews 10:24, 25*

Alaska Map
Wildflower Field Trip Route